Welcome

Thank you for choosing Page A Day Math, a great way to introduce essential math basics and writing numbers. Page A Day Math books help your child develop a solid math foundation through daily step-by-step practice, repetition, and of course, the friendly Math Squad!

How to Use This Book

1. Student traces and solves each problem, completing a page a day, front and back.
2. Parent checks answers and circles incorrect problems.
3. Student corrects errors.
4. Student colors in achievement stars each day when finished!

Have Fun

ISBN – 978-1-947286-03-0

This book belongs to _____

Dear Super Hero Math Student,

You can be a Math Squad Super Hero like Flo, Jo, Bo, Zo, and me! Practice every day and you'll be a math star too!

YOUR FRIEND,
MO

P.S. Check out what my math buddies and I are up to in the Math Squad Monthly at www.PageADayMath.com.

PAGE A DAY MATH

Count ⇨ $0 + \boxed{\equiv} = \boxed{\equiv}$

Learn ⇨ $0 + 4 = 4$

Trace ⇨ $\boxed{0} + \boxed{4} = \boxed{4}$

Copy ⇨ $\boxed{} + \boxed{} = \boxed{}$

Flo says, "Practice and be a math star like me!"

1) $0 + 4 = \boxed{}$

2) $10 + 3 = \boxed{}$

3) $3 + 8 = \boxed{}$

4) $4 + 0 = \boxed{}$

5) $7 + 3 = \boxed{}$

6) $3 + 9 = \boxed{}$

7) $4 + 0 = \boxed{}$

8) $3 + 10 = \boxed{}$

9) $0 + 4 = \boxed{}$

10) $3 + 6 = \boxed{}$

PAGE A DAY
MATH

Wow, you are learning fast. Here are a few more. Yay!

11) $0 + 4 =$

18) $3 + 4 =$

12) $10 + 3 =$

19) $4 + 0 =$

13) $3 + 2 =$

20) $7 + 3 =$

14) $4 + 0 =$

21) $3 + 5 =$

15) $3 + 9 =$

22) $8 + 3 =$

16) $1 + 3 =$

23) $0 + 4 =$

17) $6 + 3 =$

24) $3 + 0 =$

Color in the stars each day when you finish!

www.PageADayMath.com

Count ⇨ 🦴 **+** 🦴🦴🦴🦴 **=** 🦴🦴🦴🦴🦴

Learn ⇨ $1 + 4 = 5$

Trace ⇨ $1 + 4 = 5$

Copy ⇨ ☐ + ☐ = ☐

You are on your way to success! Try these!

1) $1 + 4 =$ ☐

2) $4 + 0 =$ ☐

3) $4 + 1 =$ ☐

4) $9 + 3 =$ ☐

5) $1 + 4 =$ ☐

6) $3 + 7 =$ ☐

7) $4 + 1 =$ ☐

8) $8 + 3 =$ ☐

9) $3 + 10 =$ ☐

10) $1 + 4 =$ ☐

You are coming along. Practice makes perfect!

11) $3 + 10 =$

18) $6 + 3 =$

12) $1 + 4 =$

19) $1 + 3 =$

13) $3 + 2 =$

20) $3 + 8 =$

14) $9 + 3 =$

21) $9 + 3 =$

15) $4 + 3 =$

22) $4 + 1 =$

16) $3 + 7 =$

23) $3 + 0 =$

17) $4 + 1 =$

24) $5 + 3 =$

Color in the stars each day when you finish!

DOG

Count ⇨ ▦ + ▤ = ▤

Learn ⇨ 2 + 4 = 6

Trace ⇨ 2 + 4 = 6

Copy ⇨ ☐ + ☐ = ☐

Hurray! Keep going. You've got it.

1) 4 + 2 = ☐

2) 1 + 4 = ☐

3) 2 + 4 = ☐

4) 0 + 4 = ☐

5) 4 + 2 = ☐

6) 9 + 3 = ☐

7) 2 + 4 = ☐

8) 3 + 8 = ☐

9) 4 + 2 = ☐

10) 1 + 4 = ☐

PAGE A DAY
MATH

Alright! Now review what you have learned.

11) 4 + 2 = ☐

12) 10 + 3 = ☐

13) 7 + 2 = ☐

14) 2 + 4 = ☐

15) 2 + 8 = ☐

16) 0 + 4 = ☐

17) 4 + 1 = ☐

18) 3 + 9 = ☐

19) 2 + 4 = ☐

20) 2 + 6 = ☐

21) 1 + 4 = ☐

22) 3 + 8 = ☐

23) 4 + 2 = ☐

24) 6 + 3 = ☐

☆ ☆ ☆ Color in the stars each day when you finish!

Day 4 Review

PAGE A DAY
MATH

Practice makes perfect. That's right!

1) $1 + 4 =$ ☐

2) $2 + 8 =$ ☐

3) $9 + 3 =$ ☐

4) $4 + 2 =$ ☐

5) $2 + 7 =$ ☐

6) $9 + 3 =$ ☐

7) $0 + 4 =$ ☐

8) $10 + 3 =$ ☐

9) $2 + 4 =$ ☐

10) $4 + 0 =$ ☐

11) $6 + 2 =$ ☐

12) $4 + 1 =$ ☐

13) $3 + 8 =$ ☐

14) $2 + 4 =$ ☐

© 2017 Page A Day Math, LLC

7

PAGE A DAY
MATH

Keep practicing! You are doing so well. Woof-woof!

15) $3 + 4 =$ ☐

22) $4 + 0 =$ ☐

16) $2 + 7 =$ ☐

23) $2 + 4 =$ ☐

17) $1 + 4 =$ ☐

24) $3 + 4 =$ ☐

18) $3 + 8 =$ ☐

25) $1 + 4 =$ ☐

19) $4 + 3 =$ ☐

26) $10 + 3 =$ ☐

20) $0 + 4 =$ ☐

27) $2 + 9 =$ ☐

21) $4 + 2 =$ ☐

28) $4 + 3 =$ ☐

Color in the stars each day when you finish!

PAGE A DAY MATH

Count ⇨ ⬚ + ⬚ = ⬚

Learn ⇨ 3 + 4 = 7

Trace ⇨ 3 + 4 = 7

Copy ⇨

Keep up the terrific effort. Try these!

1) 3 + 4 =

2) 4 + 2 =

3) 9 + 3 =

4) 4 + 3 =

5) 1 + 4 =

6) 2 + 4 =

7) 4 + 1 =

8) 3 + 4 =

9) 0 + 4 =

10) 4 + 2 =

PAGE A DAY
MATH

You are getting better each day. Woof! Yippee!

11) $3 + 4 =$ ☐

18) $6 + 2 =$ ☐

12) $3 + 7 =$ ☐

19) $4 + 3 =$ ☐

13) $2 + 4 =$ ☐

20) $1 + 4 =$ ☐

14) $9 + 3 =$ ☐

21) $3 + 8 =$ ☐

15) $4 + 3 =$ ☐

22) $4 + 2 =$ ☐

16) $7 + 2 =$ ☐

23) $0 + 4 =$ ☐

17) $1 + 4 =$ ☐

24) $3 + 4 =$ ☐

PAGE A DAY MATH

Count ⇨ 🦴🦴 + 🦴🦴 = 🦴🦴🦴🦴

Learn ⇨ 4 + 4 = 8

Trace ⇨ 4 + 4 = 8

Copy ⇨ ☐ + ☐ = ☐

Flo says, "Try these...woof...go for it!"

1) 4 + 4 = ☐

2) 4 + 3 = ☐

3) 1 + 4 = ☐

4) 4 + 4 = ☐

5) 2 + 4 = ☐

6) 4 + 1 = ☐

7) 4 + 4 = ☐

8) 2 + 4 = ☐

9) 4 + 3 = ☐

10) 4 + 4 = ☐

PAGE A DAY
MATH

You are on the right track. Hurray! Great effort!

11) 4 + 4 = ☐

12) 1 + 4 = ☐

13) 3 + 4 = ☐

14) 0 + 4 = ☐

15) 5 + 3 = ☐

16) 4 + 4 = ☐

17) 4 + 2 = ☐

18) 3 + 4 = ☐

19) 1 + 4 = ☐

20) 4 + 4 = ☐

21) 6 + 3 = ☐

22) 2 + 4 = ☐

23) 4 + 1 = ☐

24) 4 + 3 = ☐

Count ⇨ 🦴🦴🦴 + 🦴🦴 = 🦴🦴🦴🦴🦴

Learn ⇨ 5 + 4 = 9

Trace ⇨ 5 + 4 = 9

Copy ⇨ ☐ + ☐ = ☐

OK, now try these. Bo knows you can do it.

1) 5 + 4 = ☐

2) 4 + 3 = ☐

3) 5 + 4 = ☐

4) 4 + 2 = ☐

5) 4 + 4 = ☐

6) 1 + 4 = ☐

7) 4 + 5 = ☐

8) 4 + 4 = ☐

9) 3 + 4 = ☐

10) 4 + 5 = ☐

Terrific! Good for you. Now finish these.

11) 5 + 4 =

12) 4 + 2 =

13) 3 + 4 =

14) 1 + 4 =

15) 4 + 5 =

16) 4 + 3 =

17) 4 + 4 =

18) 4 + 1 =

19) 5 + 4 =

20) 4 + 2 =

21) 4 + 3 =

22) 4 + 4 =

23) 3 + 4 =

24) 4 + 5 =

You are working so hard. Awesome effort. Yay!

1) $5 + 4 = $ ▭

2) $4 + 1 = $ ▭

3) $9 + 3 = $ ▭

4) $0 + 4 = $ ▭

5) $4 + 4 = $ ▭

6) $4 + 5 = $ ▭

7) $3 + 4 = $ ▭

8) $4 + 4 = $ ▭

9) $4 + 0 = $ ▭

10) $4 + 3 = $ ▭

11) $5 + 4 = $ ▭

12) $8 + 3 = $ ▭

13) $1 + 4 = $ ▭

14) $3 + 6 = $ ▭

The Math Squad admires your determination. Wonderful!

15) 3 + 4 =

16) 2 + 7 =

17) 1 + 4 =

18) 3 + 8 =

19) 4 + 3 =

20) 0 + 4 =

21) 4 + 2 =

22) 4 + 0 =

23) 2 + 4 =

24) 3 + 4 =

25) 1 + 4 =

26) 10 + 3 =

27) 2 + 9 =

28) 4 + 3 =

Count ⇨ ▦ + ▦ = ▦▦

Learn ⇨ 6 + 4 = 10

Trace ⇨ 6 + 4 = 10

Copy ⇨

You are really improving. Woof-woof.

1) 6 + 4 =

2) 4 + 4 =

3) 4 + 6 =

4) 3 + 4 =

5) 5 + 4 =

6) 2 + 4 =

7) 4 + 6 =

8) 4 + 3 =

9) 4 + 5 =

10) 6 + 4 =

Super! You are doing so well. Yay!

11) 6 + 4 =

18) 0 + 4 =

12) 4 + 4 =

19) 4 + 2 =

13) 1 + 4 =

20) 4 + 5 =

14) 4 + 6 =

21) 4 + 3 =

15) 2 + 4 =

22) 6 + 4 =

16) 4 + 5 =

23) 4 + 1 =

17) 3 + 4 =

24) 4 + 4 =

Count ⇨ + =

Learn ⇨ 7 + 4 = 11

Trace ⇨ 7 + 4 = 11

Copy ⇨

You are so determined. Good for you! Arf-arf!

1) 4 + 7 =

6) 4 + 3 =

2) 6 + 4 =

7) 7 + 4 =

3) 7 + 4 =

8) 4 + 5 =

4) 4 + 4 =

9) 2 + 4 =

5) 4 + 7 =

10) 4 + 7 =

PAGE A DAY
MATH

You are improving every day. Nothing can stop you now!

11) 7 + 4 =

18) 4 + 5 =

12) 4 + 3 =

19) 6 + 4 =

13) 5 + 4 =

20) 7 + 4 =

14) 4 + 7 =

21) 4 + 1 =

15) 2 + 4 =

22) 6 + 4 =

16) 4 + 4 =

23) 0 + 4 =

17) 4 + 6 =

24) 4 + 7 =

PAGE A DAY MATH

Count ⇨ ▓▓ + ▓ = ▓▓

Learn ⇨ 8 + 4 = 12

Trace ⇨ 8 + 4 = 12

Copy ⇨

You have learned so much. Now try these.

1) 8 + 4 =

2) 4 + 7 =

3) 4 + 8 =

4) 6 + 4 =

5) 8 + 4 =

6) 5 + 4 =

7) 4 + 8 =

8) 4 + 4 =

9) 4 + 3 =

10) 8 + 4 =

You make it look easy. Way to go. You are awesome!

11) 8 + 4 = ☐

12) 5 + 4 = ☐

13) 4 + 7 = ☐

14) 4 + 8 = ☐

15) 6 + 4 = ☐

16) 4 + 2 = ☐

17) 8 + 4 = ☐

18) 3 + 4 = ☐

19) 4 + 6 = ☐

20) 4 + 8 = ☐

21) 4 + 1 = ☐

22) 4 + 4 = ☐

23) 7 + 4 = ☐

24) 4 + 5 = ☐

www.PageADayMath.com

You did it. You are a math star! Tremendous!

1) $5 + 4 =$

2) $4 + 1 =$

3) $9 + 3 =$

4) $0 + 4 =$

5) $4 + 4 =$

6) $4 + 5 =$

7) $3 + 4 =$

8) $4 + 4 =$

9) $4 + 0 =$

10) $4 + 3 =$

11) $5 + 4 =$

12) $8 + 3 =$

13) $1 + 4 =$

14) $3 + 6 =$

You have it now. Keep up the super effort. Go for it!

15) 7 + 4 =

16) 4 + 2 =

17) 4 + 5 =

18) 4 + 6 =

19) 4 + 7 =

20) 4 + 1 =

21) 6 + 4 =

22) 4 + 6 =

23) 3 + 4 =

24) 4 + 1 =

25) 7 + 4 =

26) 4 + 5 =

27) 0 + 4 =

28) 4 + 7 =

Count ⇨ $\blacksquare\blacksquare + \blacksquare = \blacksquare\blacksquare\blacksquare$

Learn ⇨ $9 + 4 = 13$

Trace ⇨ $9 + 4 = 13$

Copy ⇨

You have the hang of it. You're great at math!

1) $9 + 4 =$

2) $4 + 8 =$

3) $4 + 9 =$

4) $7 + 4 =$

5) $4 + 9 =$

6) $4 + 5 =$

7) $9 + 4 =$

8) $5 + 4 =$

9) $9 + 4 =$

10) $4 + 6 =$

Day 13

Nice going. You can be very proud of yourself. Wow!

11) 9 + 4 =

18) 4 + 8 =

12) 4 + 7 =

19) 4 + 2 =

13) 5 + 4 =

20) 9 + 4 =

14) 4 + 9 =

21) 4 + 6 =

15) 3 + 4 =

22) 1 + 4 =

16) 8 + 4 =

23) 4 + 0 =

17) 4 + 4 =

24) 4 + 9 =

PAGE A DAY MATH

Count ⇨ 🔋🔋 + 🔋 = 🔋🔋🔋

Learn ⇨ 10 + 4 = 14

Trace ⇨ 10 + 4 = 14

Copy ⇨ ☐ + ☐ = ☐

You are almost done with this book! Amazing!

1) 10 + 4 = ☐

2) 4 + 9 = ☐

3) 4 + 10 = ☐

4) 8 + 4 = ☐

5) 10 + 4 = ☐

6) 7 + 4 = ☐

7) 4 + 10 = ☐

8) 6 + 4 = ☐

9) 4 + 5 = ☐

10) 10 + 4 = ☐

PAGE A DAY MATH

Hurray! You earned a certificate! Super cool. Yippee!

11) $10 + 4 =$

18) $3 + 4 =$

12) $4 + 8 =$

19) $4 + 10 =$

13) $6 + 4 =$

20) $2 + 4 =$

14) $4 + 10 =$

21) $4 + 0 =$

15) $7 + 4 =$

22) $9 + 4 =$

16) $4 + 9 =$

23) $4 + 1 =$

17) $5 + 4 =$

24) $10 + 4 =$

PAGE A DAY
MATH

I ♥ MATH

HURRAY! YOU ARE A MATH STAR!

THE MATH SQUAD CONGRATULATES _____
FOR COMPLETING **ADDITION AND COUNTING, BOOK 4.**